YOUR KNOWLEDGE HAS VALUE

AF131199

Bibliographic information published by the German National Library:

The German National Library lists this publication in the National Bibliography; detailed bibliographic data are available on the Internet at http://dnb.dnb.de .

Imprint:

Copyright © 2017 GRIN Verlag, Open Publishing GmbH
Print and binding: Books on Demand GmbH, Norderstedt Germany
ISBN: 9783668603486

This book at GRIN:

https://www.grin.com/document/386034

William Fidler

The emergence of Grandi's series as a member of an infinite set of convergent infinite series

GRIN Publishing

GRIN - Your knowledge has value

Since its foundation in 1998, GRIN has specialized in publishing academic texts by students, college teachers and other academics as e-book and printed book. The website www.grin.com is an ideal platform for presenting term papers, final papers, scientific essays, dissertations and specialist books.

Visit us on the internet:

http://www.grin.com/

http://www.facebook.com/grincom

http://www.twitter.com/grin_com

Grandi's Series in context

Grandi's series, $S_G = 1 - 1 + 1 - 1 + 1 - \ldots \ldots \ldots$ together with its sum $S_G = 0.5$, arising naturally within an infinitely large set of convergent infinite series, representing the inverse of all the real numbers, R, in the range $1 \leq R \leq \infty$.

W M Fidler

Abstract

Grandi' series and its associated sum have been discussed extensively over many years. In this work it is shown that the series is infinite and indeed, has the sum 0.5.

The methods of solution are novel and employ accepted mathematical procedures. The first approach inverts the usual question of 'what is the solution to this problem?' to, 'of what problem is this a solution?', whilst the second uses Cesaro summation to show that in order for the sum of Grandi's series to be equal to 0.5, the series must be infinite.

Further, the existence of Grandi identities are identified and it is shown also that all of the infinite, convergent series contained within the infinitely large set may be expressed in terms of Grandi's series.

All of the functions from which the corresponding series are generated are shown to possess an indefinite number of derivatives of increasing order, each of which is expressible as an infinite series, and are hence ideal candidates for expansion in terms of Taylor's series.

Content

Introduction

The presentation of Grand's series and its sum is discussed widely in online fora in a way which is completely out of context, in that the series is presented in isolation and then wrangled over as to its sum and extent. This inevitably leads to the erroneous conclusions noted below.

At the outset it must be emphasised that the term, ' infinity' and its associated notation, ∞, do not represent a number, but rather, are shorthand for the phrase 'indefinitely large'.

Grandi's series, $S_G = 1 - 1 + 1 - 1 + 1 - 1 + \ldots \ldots \ldots$ is frequently employed to illustrate the paradoxes which may arise by the blind manipulation of infinite series. This is discussed extensively by Stewart [1].

For example, the sum of the series may be shown to be equal zero or unity by the judicious insertion of brackets in the series.

Thus, $S_G = (1 - 1) + (1 - 1) + (1 - 1) + \ldots \ldots = 0$.

Whilst, $S_G = 1 - (1 - 1) - (1 - 1) - \ldots = 1$.

Both of these results have been obtained by dubious reasoning, for, in the case where the sum is zero it has been tacitly assumed that the series contains an even number of terms, whilst the converse has been assumed to obtain the second result.

A slightly more subtle, but nonetheless incorrect conclusion is obtained by subtracting S_G from unity,

i.e. $1 - S_G = 1 - (1 - 1 + 1 - 1 + 1 - 1 + \ldots) = 1 - 1 + 1 - 1 + 1 - 1 + \ldots = S_G$, and so, $S_G = \frac{1}{2}$.

Simply because subtracting S_G from unity gives a thing which resembles S_G does not merit calling the result S_G, for it has been blithely assumed that the inclusion of another unity into the series will make no difference. At bottom, $1 - S_G = 1 - S_G$, and no more may be said.

It seems that Grandi's series is nothing more than a mathematical curiosity.

We will now show that this is anything but the case.

Consider any real number, R in the range $1 \leq R \leq \infty$.

For any of these R we may write: $R = 1 + x$, where x is a positive number in the range $0 \leq x \leq \infty$.

Further, we may form the inverse, S, of R, thus, $S = \frac{1}{R} = \frac{1}{1 + x}$. ------------(a)

Now, the quotient $\frac{1}{1+x}$ may be expanded by the simple process of algebraic division to yield: $1 - x + x^2 - x^3 + x^4 - x^5 + \dots$. Even although, in some cases the LHS may be determined with infinite precision, all of the series will extend indefinitely, unless x=0.

In compact form we then write, $S = \sum_{n=1}(-x)^{n-1}$ ---------------------- (1).

Given the form of equation (1) it is assumed that it will converge for all values of x less than, or equal to unity.

In order to obtain a series which, it is assumed, converges for $x \geq 1$ we reverse the order of the denominator in equation (a), noting that the result is still equal to S and again, expand this quotient by algebraic long division. An example of this procedure is shown on p590 of [2].

We now get : $S = \frac{1}{x} - \frac{1}{x^2} + \frac{1}{x^3} - \frac{1}{x^4} + \dots$ Remarks, similar to those above may be made, but in this case, $x = \infty$. All of this is discussed at length later in the work.

Hence, $S = \sum_{n=1}(-1)^{n-1} x^{-n}$ ------------------- (2).

We have the same expectation as that expressed above, but with the caveat that equation (2) will represent converging series if $x \geq 1$.

It will have been noticed by now that when x=1 both series become Grandi's series.

The left hand sides of both equations (1) and (2) may be determined to any accuracy for a given value of x since both are equal to $\frac{1}{1+x}$, hence it only remains to explore the right hand sides by partial summation of the terms. The convergence of these series will be decided by comparison of their partial sums with the LHSs.*

Taking x=1.265, a number chosen at random, the magnitude of the LHS of (2) is 0.4415011 by calculator. The magnitudes of the corresponding series summing the first 100 and 1000 terms are 0.44150114 and 0.44150114, respectively.*

Taking x=0.43215, again chosen at random, the magnitude of the LHS by calculator is 0.6982508. The magnitudes of the corresponding series summing the first 100 and 1000 terms are 0.698250882 and 0.698250882, respectively. For comparison, the magnitudes of the first 5 and 10 terms are 0.70877496 and 0.698092262, respectively, and show, as expected, the dependence of the summation upon the number of terms.*

<u>S = 1/2</u>

Now, we have partitioned the inverse of R in the range $1 \leq R \leq \infty$ into two regions, one where $0 \leq x \leq 1$, the other where $1 \leq x \leq \infty$.

The function $\frac{1}{1+x}$ ranges from unity to zero, passing through $\frac{1}{2}$ when x = 1, and where the associated series is Grandi's series.

We now examine convergence of both sets of series in the vicinity of x = 1; in this we denote the series given by equations (1) and (2) as S_1 and S_2 , respectively.

S_1

LHS	No. of terms	RHS
0.5025125 (x = 0.99)	10	0.0480492085
	10^2	0.318576713
	10^3	0.502490869
	10^4	0.502512563
	10^5	0.502512563
	10^6	

LHS	No. of terms	RHS
0.5002501 (x = 0.999)	10	0.00498004992
	10^2	0.0476277403
	10^3	0.316310443
	10^4	0.500227527
	10^5	0.500250125
	10^6	

LHS	No. of terms	RHS
0.500025 (x = 0.9999)	10	0.0004998005
	10^2	0.00497557943
	10^3	0.0475859325
	10^4	0.316085281
	10^5	0.500002311
	10^6	0.500025001

LHS	No. of terms	RHS
0.5000025 (x = 0.99999)	10	0.000049998
	10^2	0.00049975508
	10^3	0.00497513275
	10^4	0.0475817551
	10^5	0.316062779
	10^6	0.49997801
	10^7	0.5000025

The purpose of the above set of values is to demonstrate that, as x → 1 from above, the LHS tends towards $1/2$ and the associated series partial sums tend towards $1/2$ provided that a sufficient number of terms are included in the partial sum. The convergence of the partial sums for a set of increasing x towards their respective LHSs is taken to be evidence that the series are convergent and that as x approaches unity the number of terms in the partial sums increases; indeed, increases without limit.

From the table immediately above it is seen that, for the series to converge to the value of the LHS, requires that it extend to 10^7 terms. The time required to obtain this result on the computer available to the author was in the region of eight hours. It is estimated that if x is moved closer to unity, e.g. x = 0.999999 (S = 0.5000002) then the number of terms required for convergence could well be in the region of 10^8 and the associated computing time, $3\frac{1}{2}$ days.

Further, it follows that the number of terms in the partial sum could be extended without limit and correspondingly, S would approach 0.5.

This could be described as **'convergence from above towards S = 0.5'**.

The above analysis was also conducted on S_2. Again the same behaviour was observed, and, without displaying the plethora of tables as shown above it was found that for x = 1.00001 (S = 0.4999975) the series converged to 0.4999975 using 10^7 terms, with the concomitant computing time of eight hours.

Hence, for $x \to 1$, the behaviour of the series may be described as **'convergence from below towards S = 0.5'**

All of the foregoing is employed to establish that for x = 1, the series becomes Grandi's series with sum 0.5, and occupies a position of some import within the infinite set of convergent series representing the inverses of all of the real numbers in the range from unity to infinity, inclusive. Moreover, Grandi's series demarcates those series whose sum is less than 0.5 from those whose are greater than 0.5. It should be noted that we have not proved that Grandi's series is convergent, despite showing that its sum is 0.5. What we have shown, is that, in the convergence of series, arbitrarily close to Grandi's series, the number of terms that are required for convergence becomes arbitrarily large.*

Accordingly, we may write:

$$Lt/_{(x \to 1)} \uparrow \sum_{n=1}(-x)^{n-1} = Lt/_{(x \to 1)} \downarrow \sum_{n=1}(-1)^{n-1} x^{-n} = \frac{1}{2} \text{------------ (3)}.$$

The upper limit of the summation is implied to be infinity.

Derivatives

The function, $S = \frac{1}{(1+x)}$ is continuous in the range $0 \le x \le \infty$ by virtue of the continuity of the real numbers in this range. In addition, because of its form, S may be differentiated an arbitrary number of times. If we make the not unreasonable assumption that the corresponding series may be differentiated, term by term, then, new series may be derived.

7

As an example, consider the function, $S_1 = {}^1/_{(1+x)} = 1 - x + x^2 - x^3 + x^4 - \ldots$

$$^{dS_1}/_{dx} = S_{11} = -{}^1/_{(1+x)^2} = -1 + 2x - 3x^2 + 4x^3 \ldots\ldots\ldots (c)$$

At the 'location' of Grandi's series, $x = 1$; hence, the series, S_{G11}, which could be termed the first Grandi derivative for series 1 is: $S_{G11} = -{}^1/_4 = -1 + 2 - 3 + 4 - 5 + 6 - etc.$

Now, $S_2 = {}^1/_{(x+1)} = {}^1/_x - {}^1/_{x^2} + {}^1/_{x^3} - {}^1/_{x^4} + \ldots\ldots\ldots$

$$^{dS_2}/_{dx} = S_{21} = -{}^1/_{(x+1)^2} = -{}^1/_{x^2} + {}^2/_{x^3} - {}^3/_{x^4} + {}^4/_{x^5} \ldots\ldots(d)$$

At $x = 1$, this series becomes the first Grandi derivative for series 2, S_{G21}.

Hence, $S_{G21} = -{}^1/_4 = -1 + 2 - 3 + 4 - 5 + 6 - etc.$

There is an important caveat that must be noted at this juncture:

In view of the statements made at the beginning of this work, the temptation to introduce brackets on the RHS of either of the first Grandi derivatives must be resisted; convergence must be demonstrated by examining numerically, the behaviour of series (c) and (d) arbitrarily close to $x = 1$.

Numerical experimentation leads to the same conclusions which gave rise to equation (3).

Further numerical experimentation on 'first derivative series' reveals that, outwith the immediate vicinity of $x = 1$ (the position of Grandi's series) convergence of other series is rapid; e.g. for $x = 0.5$ the LHS of (c) is calculated to be 0.444444…, whilst the RHS converges to 0.444444.., after 35 terms. For $x = 1.2$ the LHS of (d) is calculated to be 0.20661157, whilst the RHS converges to 0.20661157 after 150 terms.*

For $x = 0$ and $x = \infty$, convergence is instantaneous, for the series contain only one term.

Series for higher derivatives may be generated and examined for convergence by the methods described above.*

Since the Grandi series may be differentiated an indefinite number of times then it follows that we may express the 'Grandi function', S_G, or any other such function, S, as a Taylor series, for such a series contains an indefinite number of derivatives of increasing order.

As is well known, the Taylor series expansion for a function $f(x)$ about a centre of convergence, a, is

$$:f(x) = f(a) + (x-a)f'(a) + {}^{(x-a)^2}/_{2!}\, f''(a) + {}^{(x-a)^3}/_{3!} f'''(a) + etc.$$

If, $x = a + h, then\ x - a = h,$ and so we have:

8

$$f(a + h) = f(a) + hf'(a) + {}^{h^2}/_{2!}\,f''(a) + {}^{h^3}/_{3!}\,f'''(a) + {}^{h^4}/_{4!}\,f^{iv}(a) + etc. \text{ --(e)}.$$

If, $x = a - h$, then $x - a = -h$, hence:

$$f(a - h) = f(a) - hf'(a) + {}^{h^2}/_{2!}\,f''(a) - {}^{h^3}/_{3!}\,f'''(a) + {}^{h^4}/_{4!}\,f^{iv}(a) + etc. \text{ --(f)}.$$

In the earlier numerical investigations the value of the function $^{1}/_{1+x}$ was easily determined (again, within the limits of the display of the calculating device) for a given value of x and the object then was to sum the corresponding series in x, to check for convergence (although, in two cases, summing 10^7 terms).

From the above expressions it may be seen that there is an easy way, not to determine the magnitude of the sum, but the difference between the value of the function at a and that at $(a \pm h)$. We illustrate this with the following example.

When $x = a = 1$ the function $^{1}/_{1+x}$ could be described as the 'Grandi function', for which the first four derivatives are: $f'(1) = -^{1}/_{4}$, $f''(1) = ^{1}/_{4}$, $f'''(1) = -^{3}/_{8}$, $f^{iv}(1) = ^{3}/_{4}$.

Hence, we rearrange equation (e) and write:

$$f(1 - h) = f(1) + (^{1}/_{4}\,h) + \left(^{1}/_{4}\,{}^{h^2}/_{2!}\right) + \left(^{3}/_{8}\,{}^{h^3}/_{3!}\right) + \left(^{3}/_{4}\,{}^{h^4}/_{4!}\right) + etc.$$

$$\therefore f(1 - h) - f(1) = (^{1}/_{4}\,h) + \left(^{1}/_{4}\,{}^{h^2}/_{2!}\right) + \left(^{3}/_{8}\,{}^{h^3}/_{3!}\right) + \left(^{3}/_{4}\,{}^{h^4}/_{4!}\right) + etc.$$

i.e. $f(1 - h) - f(1) = (^{1}/_{4}\,h) + \left({}^{h^2}/_{8}\right) + \left({}^{h^3}/_{16}\right) + \left({}^{h^4}/_{32}\right) + etc. = g(h).$ --- (g).

A formula for g(h), which, in this instance, obviates the necessity to calculate the Grandi derivatives, is: $g(h) = \sum_{n=1}^{n=\infty} {}^{h^n}/_{2^{n+1}}.$

Hence, we may write: $f(1 - h) - f(1) = \sum_{n=1}^{n=\infty} {}^{h^n}/_{2^{n+1}}.$ ---------------- (4).

Now,

$$f(1 + h) = f(1) + hf'(1) + {}^{h^2}/_{2!}\,f''(1) + {}^{h^3}/_{3!}\,f'''(1) + {}^{h^4}/_{4!}\,f^{iv}(1) + etc.$$

And, in a similar manner as above, we may write:

9

$$f(1 + h) - f(1) = \left[-(1/4\,h) + \left(h^2/8 \right) - \left(h^3/16 \right) + \left(h^4/32 \right) - etc. \right] = r(h). \text{ --(f).}$$

A formula for **r(h)**, is: $\quad r(h) = \sum_{n=1}^{n=\infty} (-h)^n / 2^{n+1}$, and so:

$$f(1 + h) - f(1) = \sum_{n=1}^{n=\infty} (-h)^n / 2^{n+1}. \text{ ----------------- (5).}$$

This process may be repeated for any **f(x)**, for all that is required is the calculation of the appropriate derivatives. A cursory examination of equations (4) and (5) show that, for small h, later terms in the series tend rapidly towards insignificance in comparison with the earlier terms—a hallmark of convergence.

Now, $dS/dx = -1/(1 + x)^2 = -S^2$; from which we conclude that $S = \pm[1/1 + x]$.

Thus, there are, in particular, two Grandi series.

The $S_G = -\dfrac{1}{2}$ corresponds to the series located at the position, $x = -1$, within the infinite set of series representing the inverse of the negative real numbers, **-R**, in the range $-1 \le -R \le -\infty$, as shown in the figure below, together with the mirror-image positive branch:

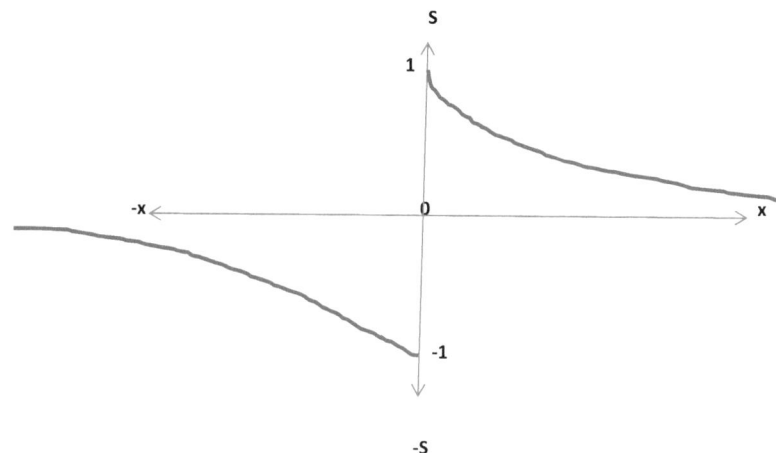

We may write the inverse, S_- of a negative real number, **-R**, in the range $-1 \le -R \le -\infty$ as:

$$S_- = 1/(-1 - x) = -1/(1 + x).$$

It then follows that the negative Grandi series is,

$$S_{G-} = -1 + 1 - 1 + 1 - 1 + 1 - etc. = {-1}/_2.$$

*The apparent rapid convergence of series outwith the immediate vicinity of $x = 1$,which are examined by numerical means in this work, is illusory, to the extent that terms in the series become so small that their effect cannot be seen as a result of the limits of the display of the calculating devices. Strictly, with the exception of the first and last series in the infinite set, viz. at $x = 0$ and $x = \infty$, respectively, which each contain only a single term and converge immediately, in any other case the series only converges at infinity, i.e. after the summation of an infinite number of terms, for whilst the partial sums may indicate convergence for a particular number of terms, none of the terms beyond this number is zero. Indeed, if we wish to be pedantic we could write the series whose sums are **1** and **0** as infinite series of the form, **1-0+0-0+0….**, and **0-0+0-0+0…** , respectively. All of the series presented here must be absolutely convergent, for unless algebraic long division is an illegimate mathematical procedure, the sum, represented by the series on the RHS, must equal the value on the LHS.

We may then argue that the veracity of the result, $S_{G+} = 0.5$, has been established beyond reasonable doubt, and it may now be said that: $1 - S_{G+} = S_{G+}$, not because the resulting series resembles Grandi's series but because $S_{G+} = 0.5$.

Since $S_{G-} = -0.5$, then, $-1 - S_{G-} = S_{G-}$, and $-1 - (-1 - S_{G-}) = S_{G-}$.

Also, $1 - (1 - S_{G+}) = S_{G+}$.Further, $1 + S_{G-} = S_{G+}$, $S_{G+} + S_{G-} = 0$,

$S_{G+}^2 + S_{G-}^2 = 0.5$, and $S_{G+} - S_{G-} = 1$.

We may even devise a Grandi complex number, as follows:

Consider the complex numbers, $z_x = P + iA$, and $z_y = B + iQ$, where $P = 1 + x$ and $Q = 1 + y$. Let $A = B = 0$. If we now take the inverses of these complex numbers and add them together, we obtain the complex number, $\overline{S_{x,y}} = S_x - iS_y$, where $S_x = {1}/_P$ and $S_y = {1}/_Q$.

When $x = y = 1$, $S_x = S_y = S_{G+}$, hence, $\overline{S_G} = S_{G+} + iS_{G-}$ (since $-S_{G+} = S_{G-}$).

Further, we now proceed to show that any series of the type examined here may be written as a multiple of Grandi' series.

Let us assume that any real number, **R**, in the range $1 \le R \le \infty$ may be written: $R = a(1 + x)$, and its inverse, $S = {1}/_{a(1 + x)}$. If we let $x = 1$, then, from either of equations (1) or (2) we get, $S = {1}/_a (1 - 1 + 1 - 1 + 1 - 1 + \cdots)$, i.e. Grandi's series multiplied by ${1}/_a$.

Now, if $x = 1$ then $a = R/2$, hence $S = {}^2/_R\ S_{G+}$. Further, we may write $R = (1 + y)$, with inverse, $S = {}^1/_{(1 + y)}$. From equation (1), we may write, $S = \sum_{n=1}(-y)^{n-1}$ for $0 \le y \le 1$

and, from equation (2), $S = \sum_{n=1}(-1)^{n-1}\,y^{-n}$ for $1 \le y \le \infty$;

but, $y = R - 1$.

Hence we have: ${}^2/_R\ S_{G+} = \sum_{n=1}(-(R-1))^{n-1}$ for $0 \le (R-1) \le 1$;

and $\qquad\qquad\qquad {}^2/_R\ S_{G+} = \sum_{n=1}(-1)^{n-1}\,(R-1)^{-n}$ for $1 \le (R-1) \le \infty$.

It is noted that the number, 2, occupies a unique status amongst the reals, and may be stated, as follows:

'When R is given by the formula, $R = (1 + x)$, the number, 2 possesses a unique status within the set of real numbers in the range, $1 \le R \le \infty$, for, it is the only number therein whose inverse may be represented solely by Grandi's series. '

Cesaro summation of Grandi's series.

Cesaro summation is well known and requires no further explanation here.

We write down a few of the Grandi partial sums together with their corresponding Cesaro partial sums as follows:

Grandi	Cesaro
1 = 1	1
0 = 1-1	1/2
1 = 1-1+1	2/3
0 = 1-1+1-1	1/2
1 = 1-1+1-1+1	3/5
0 = 1-1+1-1+1-1	1/2
1 = 1-1+1-1+1-1+1	4/7
0 = 1-1+1-1+1-1+1-1	1/2
1 = 1-1+1-1+1-1+1-1+1	5/9

These are sufficient for our purpose.

Consider the sequence of natural numbers: **1,2,3,4,5,6,…….** .

Substituting these numbers in turn into the formula: $n/(2n-1)$ gives the sequence:

1, 2/3, 3/5, 4/7, 5/9, ... which are the elements in the Cesaro sums which correspond to odd numbers of unity in the Grandi series.

Examination of the array shown above reveals that the denominators of each Cesaro sum not equal to ½ is equal to the number of unities in the corresponding Grandi partial sum.

If we describe the number of unities in any partial sum of the Grandi series as the 'penetration', d, of the the series, then, for d, odd, we may write: $2n - 1 = d$.

It then follows that we may generate any Cesaro sum,C, for an odd value of, d, from the formula: $C = (d+1)/2d$. Thus, any odd-numbered Cesaro partial sum may be generated

in terms of the penetration, d, of Grandi's series. Hence, we can generate all Cesaro partial sums, for the sum for any even number penetration is 1/2 .

From, $C = (d+1)/2d$ we calculate the following Cesaro partial sums

$$d = 1, \quad C_1 = 1; \quad d = 3, \quad C_3 = \frac{2}{3}; \quad d = 5, \quad C_5 = \frac{3}{5}; \quad d = 7, \quad C_7 = \frac{4}{7};$$

$$d = 9, \quad C_9 = \frac{5}{9}.$$

Writing $C = \frac{1}{2} + 1/2d$ shows clearly that: as $d \to \infty, C \to 1/2$.

This suggests strongly that Grandi's series is infinite with sum, $1/2$.

The convergence, or otherwise, of series and associated derivatives in the vicinity, or indeed, outwith the vicinity, of Grandi's series may, in part, be investigated further, facilitated by the BBC Basic programs appended to this work (which may be amended as necessary), but, given the time involved in actual computation, this is left to those with the computational fortitude and computing power so to do.

Addendum
On the previous page it was noted that the extent of any calculated result was constrained by the limits of the display of the calculating device. We can, in at least one instance circumvent these limitations by reference to numbers which can be written down to any desired extent.

This is illustrated by the following example:

For $x = 0.8$, $S = f(0.8) = {}^5/_9 = 0.5555555555555 \ldots \ldots ad\ inf$.

The partial sums, as far as 10^4 terms, arising from equation (1) for x = 0.8 are shown below:

Number of terms	Partial sum
10	0.495903232
20	0.549150436
30	0.55486811
40	0.55548171
50	0.555547626
60	0.555554704
70	0.555555464
80	0.555555546
90	0.555555555
10^2	0.555555555
10^3	0.555555555
10^4	0.555555555

The values in the table exhibit a characteristic feature; namely that, outwith the immediate vicinity of the 'Grandi point', the partial sums of the series rapidly approach the calculated values of S.

For $h = 0.2$ it is easy to show that, $g(h) = 0.5(10^{-1} + 10^{-2} + 10^{-3} + 10^{-4} +$ $etc, ad\ inf)$. It should be noted that it is only possible to write this result because the Grandi derivatives have been calculated previously..

Now, $f(1) = f(0.8) - g(0.2)$, which, for emphasis is written:

f(1) = 0.555555555555555555555555555555555555....ad inf

-0.055555555555555555555555555555555555... ad inf

= 0.500000000000000000000000000000000000...ad inf

Which shows that the Grandi sum, $S = 0.5$ may be also be obtained from the difference between two infinite series, for each of the numbers above are infinite series in their own right. The result only requires that the two series be aligned appropriately, as would be the case for any subtraction.

References

[1] Concepts of Modern Mathematics

   Ian Stewart

   Dover Publications Inc, 1995.

[2] Advanced Engineering Mathematics

   C R Wylie Jnr

   McGraw-Hill Book Company Inc, 1960.

BBC BASIC computer programs

```
10 INPUT NT,X

20 SA=0

30 SB = 0

40 FOR N = 1 TO NT

50   A = (-1)^(N-1)*(1/X)^N

60   B=N*((-1)^N)/(X^(N+1))

70   PRINT A,B

80   P1A = SA+A

90   P1B=SB+B

100   SA = P1A

110   SB=P1B

120 NEXT N

130 PRINT SA,  SB,  (1/(X+1)),  (-1/((X+1)^2)), (1+X)

140 REM THIS PROGRAM EVALUATES THE SUM OF THE CHOSEN NUMBER
OF TERMS (NT) FOR GIVEN VALUES OF X , WHERE X IS GREATER THAN OR
EQUAL TO UNITY, AND THE GRADIENT, WHICH SHOULD CONVERGE TO -
1/(X+1)^2 .
```

```
10 INPUT NT,X

20 SA = 0

30 SB = 0

40 FOR N = 1 TO NT

50   A = ((-X)^(N-1))

60   B = ((N+1)*((-1)^(N+1))*X^N)

70   PRINT A,B

80   P1A = SA+A

90   P1B = SB+B

100   SA = P1A

110   SB=P1B

120 NEXT N

130 PRINT SA,SB-1, (1/(X+1)), (-1/(X+1)^2), (1+X)

140 REM THIS PROGRAM EVALUATES THE SUM OF THE CHOSEN NUMBER
OF TERMS (NT) FOR GIVEN VALUES OF X , WHERE X IS LESS THAN OR
EQUAL TO UNITY, AND ALSO THE GRADIENT, WHICH SHOULD CONVERGE
TO -1/(1+X)^2
```

YOUR KNOWLEDGE HAS VALUE